A Question of Math Book

Division

by Sheila Cato
illustrations by Sami Sweeten

Carolrhoda Books, Inc./Minneapolis

This edition published in 1999 by Carolrhoda Books, Inc.

Carolrhoda Books, Inc., c/o The Lerner Publishing Group
241 First Avenue North, Minneapolis, MN 55401 U.S.A.

Website address: www. lernerbooks.com

LIBRARY OF CONGRESS CATALOGING-IN-PUBLICATION DATA
Cato, Sheila
Division / by Sheila Cato : illustrations by Sami Sweeten.
p. cm. — (A question of math book)
Summary: A group of children introduce division, using everyday examples and practice problems.
ISBN 1-57505-319-5 (alk. paper)
1. Division—Juvenile literature. [1. Division.] I. Sweeten, Sami, ill. II. Title. III. Series: Cato, Sheila, 1936- Question of math book.
QA115.C26 1999
513.2'14—dc21 98-6354

The series A Question of Math is produced by Carolrhoda Books, Inc., in cooperation with Brown Packaging Partworks Limited, London, England.
The series is based on a concept by Sidney Rosen, Ph.D.
Series consultant: Kimi Hosoume, University of California at Berkeley
Editor: Anne O'Daly
Designers: Janelle Barker and Duncan Brown

Printed in Singapore
Bound in the United States of America

1 2 3 4 5 6 - JR - 04 03 02 01 00 99

Meet Mia. This is her dog, Popcorn. Mia is learning about math the fun way with some questions about division. Digit is here to help her work out the answers to her problems.

Math is all around us – at home, at school, in the park, at the seaside. We use math all the time to help us work things out. You can join in with Mia and Digit to find the answers to Mia's questions. You could use buttons, marbles, pennies, balls, and beads to help you work out the problems. Have fun!

It's a sunny day. Brad and I are going to the beach. We'll wear hats to protect our heads from the sun. I have 2 hats. How can I share them with Brad?

Let Brad choose the hat he wants to wear. Now you take the other hat. You have shared 2 hats between 2 people. Each person gets 1 hat. When we share something with 1 or more people, and each person gets the same amount, we call it division.

We can write about division using special signs and numbers. We say "2 divided by 2 equals 1" and we can write it as

$$2 \div 2 = 1$$

This is called a division equation. The first 2 is the number of hats to be shared. The next 2 is the number of people sharing the hats. The 1 is the number of hats you each get.

Thanks, Digit. Now I know how to write an equation about sharing 2 things.

Special Signs

Just as we have road signs to tell us what to do, we have signs in math too. The sign ÷ means divided by. The sign = means equals.

Now You Try

You have 2 apples to share with a friend. Use the special signs and numbers to show how many apples you each have.

Here I am with Josh. My mom has given me enough money to buy 4 ice cream cones. I'm going to get 2 mint cones and 2 strawberry cones. How many ice cream cones will we have each?

Well, Mia, you can give them out 1 at a time. Let Josh choose the first ice cream cone and you take a cone. Now it's Josh's turn to choose another ice cream cone and you have the last cone.

You've shared 4 ice cream cones between 2 people and you each have 2 cones. We can write a division equation for this

$$4 \div 2 = 2$$

So when I'm dividing something, division helps me find out the size of each share. Thanks Digit, I'll remember that!

Number Sentence

When you write a story, you put the words into sentences. You can use words as well as numbers and signs to write about division. This makes a number sentence. "Four divided by two equals two" is a number sentence.

Now You Try

If Mia had enough money for 6 ice cream cones, how many ice cream cones would she and Josh each have? Use beads or buttons to help figure this out.

I'm going on a picnic with 3 friends, so there will be 4 of us. I know I'll have 12 muffins to share. How can I share the muffins so that everyone gets the same number?

Think about how many muffins you have to share and how many people need to have the muffins. One way of doing this is to give a muffin to each person. Keep on handing the muffins out until there are no muffins left. Each person has 3 muffins. Everyone has the same number of muffins. There are no muffins left over.

They would each have 3 ice cream cones

Fair Shares

Knowing how to divide means that all the friends at the picnic get a fair share. Everyone gets the same amount. No one is left with no cakes to eat!

But there's another way to share the muffins. You can use the equation

$$12 \div 4 = 3$$

So if I know the answer to this equation, I know that everyone gets 3 muffins!

Now You Try

If you had 8 muffins to share between 4 people, how many would each person get? Think of a way to share the muffins. Think of how you would write this as an equation.

Popcorn has 8 puppies. They've grown too big for their bed, so I'm going to put them into 2 boxes. How many puppies will there be in each box?

We can work that out! This time we're going to share things between boxes instead of between people. It's still division. Let's start by choosing one of the puppies and putting it in a box. Then take another puppy and put it in the other box. That leaves 6 puppies. Put the puppies one at a time into the boxes in turn. Keep on doing this until the basket is empty.

They each get 2 muffins. $8 \div 4 = 2$

I started with 8 puppies in a basket. Now there are 4 puppies in each box. 8 puppies shared between 2 boxes equals 4 puppies in each box. I can write this as an equation

$$8 \div 2 = 4$$

Now I know that 8 divided by 2 equals 4. Next time, I can separate the puppies into their boxes quickly!

Sharing

Division is about sharing numbers of things. The things are shared into equal groups. It doesn't matter what you're sharing. The things you share can be hats, ice cream cones, muffins, or puppies.

Now You Try

What if Popcorn had 6 puppies? Write an equation to show how they could be divided into 2 boxes.

I know that division means sharing things so that everyone gets the same amount. How can I share 10 things into 2 equal groups?

When we talk about equal amounts, we use the word set or group. You want to separate a set of 10 into 2 equal sets. That's really easy. Remember those finger puppets you got for your birthday? Put the puppets on your fingers and hold your hands up. Look carefully at both hands. Does this help you find the answer?

$6 \div 2 = 3$

Of course, Digit. My hands are 2 equal sets of 5. I have 5 puppets on each hand. To show that 10 things can be divided into 2 sets of 5, I can write

$$10 \div 2 = 5$$

And I can use my fingers to help me work out the numbers up to 10.

Even Numbers

Numbers that end in 0, 2, 4, 6, or 8 are called even numbers. 2, 4, 6, 8, 10, 12, 14, 16, 18, 20, and so on are even numbers. All even numbers can be divided by 2.

$2 \div 2 = 1$
$4 \div 2 = 2$
$6 \div 2 = 3$
$8 \div 2 = 4$
$10 \div 2 = 5$

Can you see the pattern?

Now You Try

Can you think of another set of 10 that makes 2 equal sets of 5? (The set belongs to you!)

I'm having a party today. I have 10 balloons and I want to give them to my friends so that each person has 2 balloons. How many people can I give them to?

That's like saying 10 divided by "something" equals 2. And you want to find what the something is. Can you think of a way to do this?

I know I have 10 balloons and they will be divided into sets of 2. I could take the 10 balloons and give 2 balloons to each person until I run out of balloons.

You have 10 toes. Each foot has a set of 5 toes

How Many Can Have a Share?

Sometimes you know how many things are in the set that is being shared and you know how big the share is. But you want to find out how many will get a share. Separate the big set into smaller sets. Then count the number of smaller sets. This is another way of using division.

So now all I have to do is count how many people are holding balloons. There are 10 balloons. I gave 2 balloons to each person and there are 5 people holding balloons. I can write this as a division equation

$$10 \div 5 = 2$$

We'd better hold on tight to the strings so our balloons don't fly away!

Now You Try

2 of the balloons burst, so there are 8 to share. How many people can hold 2 balloons each?

I'm going to the park with Josh and Luis. The wind is blowing and it's a great day to fly a kite. I'm taking my 3 kites to share with my friends. I'm not sure how to write a division equation about this. Can you help me?

To write the equation, start with the number of things you want to share – 3 kites. Now divide that number by the number of people who will be using them – 3 children.

4 people can each hold 2 balloons

Dividing by the Same Number

When you divide any number by itself, the answer is always 1.

$1 \div 1 = 1$

$2 \div 2 = 1$

$3 \div 3 = 1$

Divide the biggest number you can think of by itself. What is the answer?

When you share 3 things between 3 people, each person gets 1 thing. I didn't know you can have a set with only 1 thing in it. I can write a division equation for this

$$3 \div 3 = 1$$

We each have 1 kite. Make sure the strings don't get twisted!

Now You Try

If Holly arrives at the park with her kite and joins the others, how will the 4 kites be shared? Write an equation to show how.

Josh, Luis, and I are at the skating rink. We have just enough money to rent 6 skates. Will that be the right number of skates for 3 people to have a pair?

This time we'll use a different way to share things. You can't skate with only 1 skate. You need 2 skates – 1 for each foot. 2 of something is called a pair. Let Josh take a pair first. That leaves 4 skates. Next Luis can take a pair. How many skates are left?

4 children and 4 kites. $4 \div 4 = 1$

Making Sure

Here's a way to check your answer. When you have divided a set into smaller sets, add together all the things in the smaller sets. You should have the number of things in the big set. Mia divided 6 skates into 3 sets of 2 skates. $6 \div 3 = 2$ and $2 + 2 + 2 = 6$.

Okay, Digit, I can work out the rest of the problem. We started with 6 skates. Josh took a pair and that left 4 skates. Luis took a pair and that left 2 skates. The last 2 skates are for me. So the equation is

$$6 \div 2 = 3$$

6 is the right number of skates for Josh, Luis, and me to have a pair each.

Now You Try

How many pairs of skates are there in 10 skates? Remember that a pair is 2 things.

My bedroom is a mess because my toys are all over the place. I'm going to clean my room by putting my toys away. I have 9 toys and there are 3 shelves. I'll put the same number of toys on each shelf. What's the quickest way to do this?

You could take the toys one at a time and put them onto the shelves, just like you did with the puppies. Now that you understand division so well, we can work a little faster. Choose 3 toys and put 1 on each shelf. 3 toys from 9 toys leaves 6 toys.

There are 5 pairs of skates

Take the next 3 toys and put 1 on each shelf. There are 2 toys on each shelf now. How many toys are left? What will you do with them?

I have 3 toys left and I'll put 1 more on each shelf. We started with 9 toys and put them into groups of 3 on 3 shelves.

$$9 \div 3 = 3$$

So there are 3 toys on each shelf and I have a clean bedroom!

Equal Groups

Division is about putting things into equal groups. You can do this by sharing things out, one by one, like the puppies. You can also do this by splitting a big set into smaller sets and sharing them out, like the toys.

Now You Try

If Mia had 12 toys to put into 3 equal groups, how many toys would there be in each group?

In our classroom there's a big fish tank with 15 fish in it. The fish are all squashed together. They look really sad. We're going to put them into 5 smaller tanks with the same number of fish in each tank. How many fish go into each of the small tanks?

You want to divide 15 fish into 5 equal groups. Remember how you cleaned your room? You can divide the fish in the same way.

Okay, Digit, this is what I'll do. First I'll take 5 fish out of the big tank. Next I'll put 1 of these fish into each of the 5 small tanks. That leaves 10 fish in the big tank. I'll keep on taking the fish out of the big tank 5 at a time until the big tank is empty.

Now the fish are all in the small tanks. There are 3 fish in each tank, so 15 divided by 5 equals 3. The division equation for this is

$$15 \div 5 = 3$$

The fish look much happier now that they have more room to swim around.

That's Odd!

When we say a number is odd, it doesn't mean the number is strange. Odd numbers end in 1, 3, 5, 7, and 9. Odd numbers can only be divided by other odd numbers. The answer is always another odd number!

Now You Try

Write down all the odd numbers between 1 and 10. Which of these numbers can be divided by 5?

1, 3, 5, 7, and 9. Only 5 can be divided by 5

I'm having a barbecue in my backyard. My dad has cooked 15 hot dogs for Holly, Josh, Brad, Luis, and me. He says we can have 3 hot dogs each. Will there be enough hot dogs to go around?

Here's a way to divide the hot dogs. Ask your friends to stand in line. You're at the front of the line, so you get the first 3 hot dogs. Brad is next in line, so he gets the next 3 hot dogs. Holly is behind Brad, so she gets the next 3 hot dogs. How many hot dogs are left?

I've got 3 hot dogs, Brad has 3, and Holly has 3. We've taken 9 hot dogs. That leaves 6 – 3 for Josh and 3 for Luis. If there are 15 hot dogs, 5 people can have 3 hot dogs each.

$$15 \div 3 = 5$$

So you all get your hot dogs!

Keep Taking Away

Division is a quick way of taking away lots of times. When Mia and Digit divided the hot dogs, they kept taking away a set of 3 until there were none left.

$15 - 3 - 3 - 3 - 3 - 3 = 0$

This is another way of saying that there are 5 sets of 3 in 15.

Now You Try

If Mia's father cooked 20 hot dogs, would there be enough for the 5 children to have 4 hot dogs each? Write a division equation for this problem.

We're going to the zoo. I've got 12 bags of peanuts to give to the animals. I want to share the bags with Holly and Brad. How can I do this so we each have the same number of bags?

You have 12 bags of peanuts and there are 3 children. So you want to know how many sets of 3 there are in 12. Take 3 of the bags and give 1 to each person. How many bags are left? What will you do next?

Yes, there would be enough hot dogs. The division equation looks like this $20 \div 5 = 4$

When I've given 1 bag to each person, there will be 9 bags left. Then I'll take another 3 bags and hand them out. We'll have 2 bags each. Then I'll share out the rest. So 12 bags of peanuts shared by 3 people means 4 bags each. I can write that as a division equation

$$12 \div 3 = 4$$

Thanks, Digit. You've helped me see that division is really easy.

Division Has an Opposite

Did you know that division has an opposite? Division is about separating sets. Another type of math, called multiplication, is about joining sets. Multiplication is the OPPOSITE of division.

Now You Try

Suppose the 3 children have 18 bags of peanuts to share. How many would each person get?

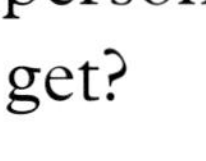

Luis and I are picking apples from the tree in my garden. We've picked 13 apples and now we want to eat them. How can we share the apples between us?

When you have more than 10 things to share, you can separate the number into smaller sets. 13 is the same as 10 plus 3. First you share 10 between you, then you share the other 3.

They had 6 bags of peanuts

I already know that 10 divided by 2 gives 5. I'll take 5 apples and give 5 to Luis. Now I can divide the 3 that are left. If we each take 1 apple, that's 2 apples. But there's 1 apple left over.

Some numbers won't divide into equal groups. There's something left over at the end. The thing that is left over is called the remainder. We can write the division equation like this

$$13 \div 2 = 6, \text{ remainder } 1$$

And Popcorn can eat the apple that's left over. I hope she doesn't get a stomachache!

Remainders

Mia had 1 apple left over because 13 is an odd number. When odd numbers are divided by 2, there is always 1 left over. This is called a remainder.

$3 \div 2 = 1$, remainder 1
$5 \div 2 = 2$, remainder 1
$7 \div 2 = 3$, remainder 1

Can you see the pattern?

Now You Try

How would you divide 10 apples between 3 people? Use buttons to help you figure it out.

My last question is about my favorite food – pizza. This pizza is too big for me to eat on my own. How can I share it with Holly?

Your pizza is a whole or a single unit, so you can't divide it like you did the apples or the bags of peanuts. You have to cut it into 2 equal pieces. Cut the pizza straight across the middle. You now have 2 equal parts. Each part is called a half.

Each person gets 3 apples. There is 1 apple left over. $10 \div 3 = 3$, remainder 1

If I want to share my pizza with 3 friends, I need to cut it into 4 equal pieces. Right?

That's right, Mia. When you cut something into 4 equal pieces, each piece is called a quarter. A quarter is a type of fraction.

So now everyone gets a piece of pizza!

Fractions

When you cut a whole thing into equal pieces, each piece is called a fraction. Knowing about fractions helps you share things.

Now I know that division is not just numbers and signs. It is about sharing things so that everyone gets the same. It is about finding the size of the share. Division helps me see how many people I can share with. I know what to do if the things I'm sharing don't divide equally. And I can even share my pizza with my friends!

Here are some useful division words.

Equation: An equation is like a sentence that we use to write about math. It has numbers and special signs instead of words.

Fraction: When you cut something into pieces that are all the same size, each piece is a fraction.

Multiplication: This is a type of math that joins numbers together. It is the opposite of division.

Pair: A pair is two of the same thing. Our hands and feet come in pairs.

Remainder: Sometimes when you divide a number there's something left over at the end. This left-over part is called a remainder.

Sets: Sets are groups that contain the same numbers of things.